W O R L D   O F   W O N D E R

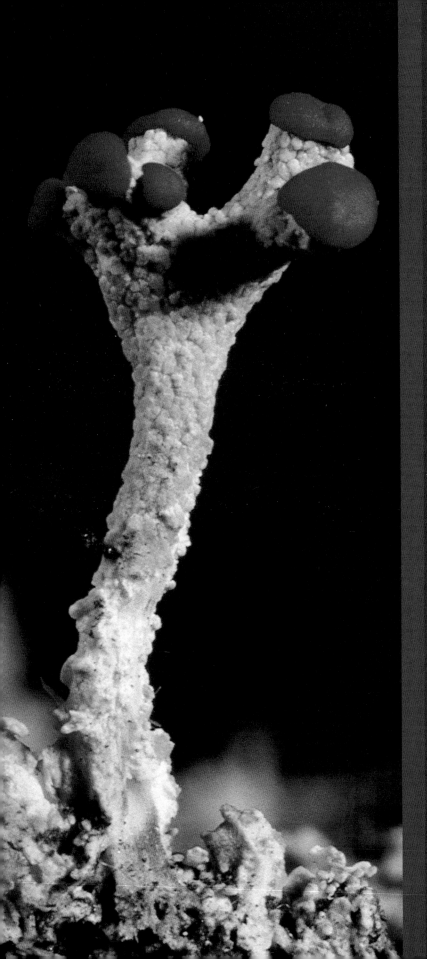

Published by Creative Education
123 South Broad Street
Mankato, Minnesota 56001

Creative Education is an imprint of
The Creative Company.

Art direction by Rita Marshall
Design by The Design Lab
Photographs by Affordable Photo Stock (Francis E. Caldwell),
Corbis (Brandon D. Cole, Ecoscene, Darrell Gulin, Peter
Johnson, Robert Landau, Frank Lane Picture Agency, Jon
Sparks), The Image Finders (Alan Chapman), JLM Visuals
(J.C. Cokendolpher, Richard P. Jacobs, Breck P. Kent), James
P. Rowan, Randal D. Sanders, Seapics.com (Doug Perrine),
Tom Stack & Associates (Sharon Gerig, John Gerlach, Milton
Rand, Doug Sokell, Inga Spence), Visuals Unlimited (Inga
Spence)

Library of Congress Cataloging-in-Publication Data

Hoff, Mary King.
Living together / by Mary Hoff.
p. cm. – (World of wonder)
Summary: Describes various examples of symbiosis, the elaborate give-and-take of food, shelter, and other essentials of
life that goes on between different kinds of animals and
plants.
ISBN 1-58341-236-0
1. Symbiosis–Juvenile literature. [1. Symbiosis.] I. Title.

QH548 .H64 2002
577.8'5 – dc21   2001047885

First Edition

9 8 7 6 5 4 3 2 1

cover & page 1: oxpeckers on a buffalo
page 2: red crest lichen
page 3: wasp pupae on a worm

Creative Education presents

WORLD OF WONDER

LIVING TOGETHER

BY MARY HOFF

Ants that grow fungi in miniature farms 🌿 Worms that live in

other animals' brains ✳ Fish that use bacteria as flashlights

🐁 The world is full of creatures that live in close relation-

ships with other kinds of living creatures. This type of rela-

tionship is called a **symbiosis**.

SYMBIOSES ARE ALL AROUND US. From the tops of mountains to the

depths of the ocean—and even inside our own bodies—many living things

are involved in this elaborate give-and-take of food, shelter, and other

essentials of life.

A red-billed oxpecker grooms a zebra

ANT FARMS

Scurrying to and from their nests in the forests of Central and South America, attine ants gather tiny pieces torn from leaves and flowers into a pile. The ants chew and scratch at the material and add drops of liquid manure from their own bodies onto the pile. Then they bring in bits of fungus and plant them in the "garden" they have made.

NATURE NOTE: *The word symbiosis comes from the Greek terms* syn, *meaning "together,"* and bios, *meaning "life." The plural of symbiosis is symbioses.*

The fungi feed on the plant material. They get the **nitrogen** they need to grow from the liquids supplied by the ants. The fungi also break down the cellulose, a tough organic material, in the plants to make a kind of food that the ants can eat. Both the fungi and the ants benefit from the relationship. This kind of "win-win" symbiosis is known as **mutualism**.

Attine ants are not the only ants that farm. The dairying ant, which is found on every continent but Antarctica, is

Dairying ants give aphids food and protection

like a dairy farmer who raises cows to make milk. Instead of milking cows, however, this ant milks tiny insects called aphids.

Dairying ants place aphids on plants. The aphids eat the plants and produce a sweet liquid called honeydew. When the ants touch the aphids with their antennae, the aphids feed honeydew to the ants. The aphids benefit from this relationship because the ants help them find a food source. They also gain protection from other insects that avoid the ants.

SYMBIOTIC SANDWICH

Many boulders and tree trunks have crusty blotches of orange, green, or brown. These growths, called lichens, are actually two kinds of organisms—algae and fungi—sandwiched together in a symbiosis so close that they look like just one living thing. Some lichens are flat. Others grow like tiny plants on the ground or hang from the branches of trees.

NATURE NOTE: *There are about 13,500 kinds of lichens found around the world.*

Lichens are a close union of two organisms

The algae and fungi that form lichens each provide something the other doesn't have. The algae contain **chloro-phyll**, a pigment that helps them capture the sun's energy and turn it into food in a process called photosynthesis. They also contribute vitamins. The fungi share the moisture and nutrients they absorb from their surroundings.

Other kinds of algae form a symbiosis with coral polyps, the tiny, ocean-dwelling animals that make coral reefs in shallow

These trees are covered by moss and lichens

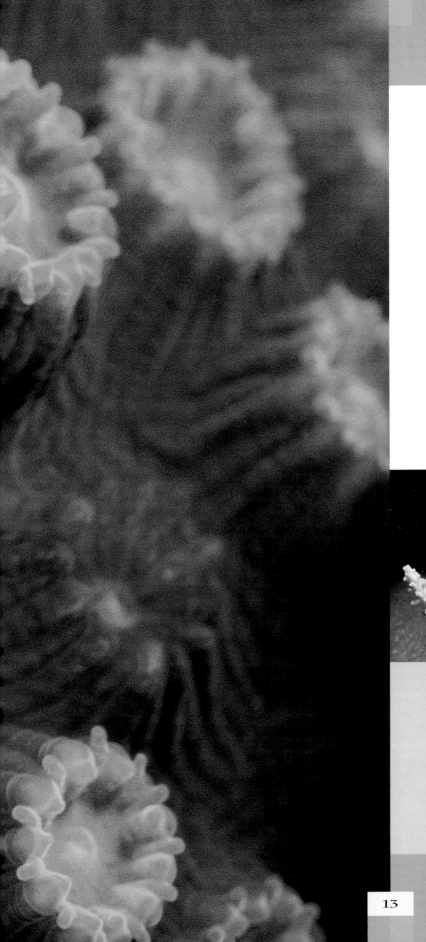

waters. The algae live inside the corals' bodies and share the energy they capture from the sun with the corals. The algae also give the corals their color. The corals, meanwhile, provide the algae with a place to live and access to light.

NATURE NOTE: *Pollution can kill the symbiotic algae that live within some corals, causing the corals to turn white.*

Coral polyps join with algae in shallow waters

INSIDE STORY

Symbiosis happens inside of people, too. Billions of tiny bacteria live in our intestines. They help our bodies to grow and stay strong by breaking down our food and providing vitamins. We help them by feeding them every time we eat.

Many **herbivores** have special bacteria in their digestive tracts that help break down the tough parts of plants and make vitamins and other compounds the animals need. Some herbivores called ruminants have four-chambered stomachs where these bacteria do their jobs. These animals shelter their bacterial buddies and provide food for them in the form of grass or other plant material. They let the bacteria work on the food in the stomach for a while, then pass the food back up into their mouths and chew it some more to make it more digestible. This process is known as rumination or cud-chewing.

NATURE NOTE: *Ruminants (cud-chewing animals) include cows, sheep, antelope, camels, deer, and giraffes.*

15 *Cattle share their food with digestive bacteria*

PLANT PARTNERS

Beneath the surface of the earth, the roots of a bean plant grow through the darkness. As they do, they give off chemicals that send a signal to soil bacteria called **rhizobia**. The rhizobia then give off a substance that makes the roots grow lumpy living-places called nodules. The rhizobia work their way into the nodules, where they find food and shelter.

NATURE NOTE: *Most of the nitrogen that exists in forms that can be used by plants and animals on land was taken from the air by rhizobia.*

Why would a plant encourage bacteria to invade its roots? Because the rhizobia turn nitrogen gas, which is found in the atmosphere, into a form of nitrogen the plants can use to make the proteins they need to live and grow.

Another important symbiotic relationship that takes place in the earth are connections, called **mycorrhizae**, between the roots of certain plants and fungi. The fungi help provide nutrients to the plants, and the plants share the food they make through photosynthesis with the fungi.

TRASH AND TROUBLE

Some symbioses involve one animal making food out of another animal's trash or trouble. An African bird called the oxpecker spends much of its life riding on the backs of rhinos, zebras, and other large mammals. The oxpecker eats ticks and flies that bug its four-legged friends. It gets a meal while the larger animal gets groomed.

A coral called *Oculina arbuscula,* found in the Atlantic Ocean off the east coast of the United States, shelters an animal called a clinging crab. The crab helps the coral by eating seaweeds that try to grow on it. This helps keep the coral healthy. The crab benefits because it can hide from its enemies amid the coral's "branches" and eat mucus secreted by the coral.

NATURE NOTE: *Oxpeckers are not entirely helpful. In addition to picking pests off of the animals they ride, they sometimes open up scabs and snitch a little blood from the animal.*

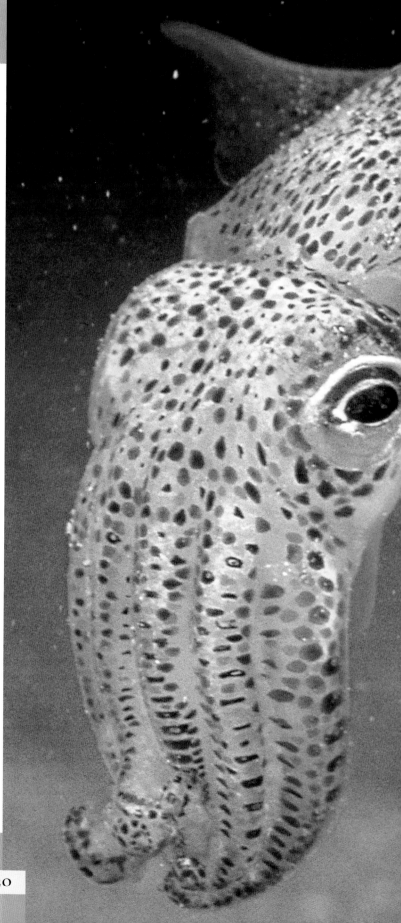

FLASHLIGHTS OF THE SEA

In the Pacific Ocean off the coast of Hawaii swims an animal called the bobtail squid, or *Euprymna scolopes*. This little animal has two indentations on its body that are filled with millions of luminescent (light-producing) bacteria. These "pockets" are lined with cells that act as reflectors. The bacteria get food, oxygen, and shelter from the squid. The light the bacteria give off helps protect the squid by confusing animals that otherwise might eat it.

⬩ Some species of ocean fish have "lights" powered by bacteria, too. These "flashlight fish" use the light the bacteria provide to help them see, evade predators, find food, and communicate. One kind of flashlight fish has a flap of flesh near its light organs that it can use to cover and uncover its supply of glowing bacteria, in effect turning its light on or off.

NATURE NOTE: *Bobtail squids don't have glowing bacteria when they hatch. They gather the bacteria from the water.*

Glowing bacteria help protect the bobtail squid

JUST HANGING AROUND

Many kinds of symbiotic relationships benefit both organisms involved. Sometimes, however, one benefits, and the other is neither helped nor harmed. This type of symbiosis is called **commensalism**.

An example of commensalism is plants that grow on other plants. Such plants are common in rain forests, where a ceiling of tall trees makes it hard for a plant to get light if it starts its growth on the ground. These plants are called **epiphytes**. Spanish moss, which hangs down from the branches of trees, is a kind of epiphyte. Other epiphytes include licorice fern, dodder, some orchids, and hanging basket moss.

Epiphytes need no soil, growing high in trees

UNWANTED GUESTS

Some plants not only grow on other plants, but also take water, food, and minerals from them without giving anything in return. These plants are involved in a third kind of symbiosis called **parasitism**. In this kind of symbiosis, one organism benefits at the other's expense. The organism that benefits is called a parasite, and the one

NATURE NOTE: *Lichens are very hardy. Some can survive temperatures of -400 °F (-240 °C)!*

This lichen is known as "old man's beard"

that provides the benefits and is harmed is called a host.

Mistletoe, a plant sometimes used to decorate homes during the holidays, is a parasite that grows on trees. Birds spread it from one tree to another when they eat the berries. Another plant parasite, called *Rafflesia arnoldii*, grows on the islands of Borneo and Sumatra in Asia. Rather than making its

NATURE NOTE: *The plant parasite rafflesia grows huge flowers. A single flower may grow to be five feet (1.5 m) across!*

Mistletoe can overwhelm trees and kill them

own food through photosynthesis, rafflesia sends out threads of tissue into other plants. Then it sucks the plants' food through these threads.

✍ Many animals have parasites living on or inside them. The worms that sometimes live in the digestive systems of dogs and cats are one common kind of parasite. People can get worms, too. They can rob us of food and energy we need to stay healthy. Veterinarians and doctors prescribe special medicine to get rid of these parasites.

Pets such as cats may have worm parasites

NATURE NOTE: *Parasites can have parasites, too. For example, fleas are hosts to tiny creatures called protozoa.*

The brain worm is a kind of worm that infects white-tailed deer when they browse on plants in which snails that carry the worm are hiding. The brain worm climbs from the intestines up the spinal column and into the deer's brain. Heartworms grow in the hearts of dogs and some other animals. When they are immature, heartworms live in the animals' blood. The parasites spread when mosquitoes carry heartworm-harboring blood from one animal to another.

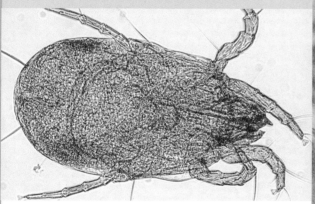

THE FABRIC OF LIFE

These are just some examples of the many symbiotic relationships around the world. From microscopic algae making food for fungi, to birds cleaning the backs of zebras, to huge tape-

Algae in a symbiosis with salamander eggs

worms sharing meals with whales, our planet is filled with symbiotic creatures. Like multicolored threads, these creatures interweave with many others to create the bright and beautiful fabric of life on Earth.

✤ Symbiosis is one special way living things depend on each other. But the more we learn about nature, the more we can see that every creature is part of a bigger picture. Whenever we affect one living thing, we affect many others. By considering the impact our actions have on the environment and its wild creatures, we can help ensure the future health and beauty of this amazing world, this world of wonder.

NATURE NOTE: *Tapeworm parasites found in whales can grow to be 100 feet (30 m) long!*

WORDS TO KNOW

Chlorophyll *is a pigment found in plants that helps them capture energy from the sun and turn it into food.*

The type of symbiosis in which one organism benefits and the other is neither helped nor hurt is called **commensalism**.

Plants that grow on other plants are called **epiphytes**.

Herbivores *are animals that eat plants.*

The type of symbiosis in which all organisms involved benefit is called **mutualism**.

Mycorrhizae *are symbiotic relationships between fungi and plant roots.*

Nitrogen *is a nutrient used by living things to make proteins.*

The type of symbiosis in which one organism (the parasite) benefits at the expense of another (the host) is called **parasitism**.

Rhizobia *are bacteria found in the soil that help turn nitrogen from the air into a form plants can use.*

A **symbiosis** *is a close relationship between different kinds of living things.*

INDEX